Tons of Trucks!

CONCRETE MIXERS

By Luke Harasymiw

Gareth Stevens PUBLISHING

Please visit our website, www.garethstevens.com. For a free color catalog of all our high-quality books, call toll free 1-800-542-2595 or fax 1-877-542-2596.

Library of Congress Cataloging-in-Publication Data

Names: Harasymiw, Luke, author.
Title: Concrete mixers / Luke Harasymiw.
Description: New York : Gareth Stevens Publishing, [2020] | Series: Tons of trucks! | Includes index.
Identifiers: LCCN 2019009900| ISBN 9781538245057 (pbk.) | ISBN 9781538245071 (library bound) | ISBN 9781538245064 (6 pack)
Subjects: LCSH: Concrete mixers–Juvenile literature.
Classification: LCC TA439 .H297 2020 | DDC 624.1/834–dc23
LC record available at https://lccn.loc.gov/2019009900

First Edition

Published in 2020 by
Gareth Stevens Publishing
111 East 14th Street, Suite 349
New York, NY 10003

Designer: Katelyn E. Reynolds
Editor: Therese Shea

Photo credits: Cover, p. 1 Blanscape/Shutterstock.com; p. 5 Ranglen/Shutterstock.com; p. 7 supakrit tirayasupasin/Shutterstock.com; pp. 9, 13, 24 (drum) Aisyaqilumaranas/Shutterstock.com; pp. 11, 24 (concrete) Nopphinan/Shutterstock.com; p. 15 Robert Pernell/Shutterstock.com; p. 17 Eng. Bilal Izaddin/Shutterstock.com; p. 19 DAMRONG RATTANAPONG/Shutterstock.com; p. 21 Chad Zuber/Shutterstock.com; p. 23 Gala_Kan/Shutterstock.com.

Printed in the United States of America

CPSIA compliance information: Batch #CW20GS: For further information contact Gareth Stevens, New York, New York at 1-800-542-2595.

Contents

My Favorite Truck 4
The Drum 8
On the Job 14
Useful Truck 22
Words to Know 24
Index. 24

This is my toy truck.
It's a concrete mixer!

This is a real one!
It's a big truck.

It has a round part.
That's called a drum.

70 80
HOTLINE SPAD
1-800-88-SPAD (7723)

The drum holds concrete.
Concrete is used
to make things.

A motor makes
the drum spin.
The drum
mixes concrete.

The truck carries
the concrete.
It goes where
it's needed.

Wet concrete pours out.

It's spread out.

061

It dries.
It makes a road!

What a useful truck!
I love trucks!

Words to Know

concrete

drum

Index

drum 8, 10, 12

mix 12

road 20

spin 12